WE CAN HELP

Tonny Rutakirwa
Illustrations by: Rica Cabrex

Published by Tonniez Publishing Press,
a member of Tonniez Group of Companies
Copyright © Tonny Rutakirwa 2020

All rights reserved. No part of this publication may be reproduced in any form whatsoever without written permission from the copyright holder or publisher.

c/o Tonniez Group Holdings,
86-90 Paul Street, London, EC2A 4NE

Emails: books@rutakirwa.com

Websites: https://www.TonnyRutakirwa.com
https://www.Rutakirwa.com
https://www.TonniezGroup.com

ISBN (eBook): 978-1-716-40730-7
ISBN (Paperback): 979-8-631-51724-0
ISBN (Paperback): 978-1-716-40728-4
ISBN (Hardback): 978-1-716-40726-0

This book belongs to:

_____

One day, an Arctic tern brought Pecko a message from his friend.

'Dear Pecko, I have not seen you for a very long time and the ice is all melting – I don't know what to do. I wonder, can you come and visit me? I would like that very much. All the best, Polo.'

Polo was a polar bear and he lived very far away from Pecko's house. 'I must take some supplies with me,' thought Pecko, and he went to fetch his favourite backpack. He thought for a minute and then put in some food; his favourite, dried fish – yum!

He also put in some scissors, some rope, some bandages, and a few other odds and ends. "I think that's enough," Pecko said to himself.

"Bye, Mum. Bye, Dad," said Pecko as he headed off. Mum and Dad watched him go, waving all the while.

A penguin travelling alone can go very fast and Pecko was soon far from home. He missed his home and his mum and dad, but he was also excited to be travelling. He couldn't wait to see his friend again. He was just thinking about the time that he and Polo had… But what was that noise?

"Help! Oh, won't somebody please help?!" The voice sounded muffled, as though the speaker was trying to talk through a closed mouth.

Which they were, Pecko soon found out! It was an enormous turtle, one of the biggest Pecko had ever seen.

"I've heard about you!" said Pecko, amazed. "You're a leatherback turtle and you're as big as a car!"

"I know I'm a leatherback!" muttered the turtle grumpily. "And what's a car?"

Pecko didn't know either, he'd just heard leatherbacks described that way! Pecko could see that a big fine net had somehow got wound around the leatherback's head and front flippers, stopping her from swimming properly and almost tying her mouth shut.

"Can you help me, or are you just going to stare?" grumbled the leatherback, still grumpy.

Pecko thought that it must be horrible having something tied around your face like that, so he didn't get upset. "I can help!" he said, taking off his backpack.

He took out his scissors and, in a flash, he snipped through the net. It fell away from the leatherback who stretched her flippers happily. "Ahhh!" she roared, stretching her mouth as wide open as possible (and that was pretty wide!). "That's better! These horrible nets are everywhere these days.

I remember when the sea was clean and beautiful... Which way are you going? I'll give you a ride to say thanks!"

"Thank you!" said Pecko. "But give me a minute."

Pecko gathered up all the pieces of net and ran to put them in a bin nearby. The leatherback nodded her approval when Pecko came back. Then Pecko hopped on her broad back and whoosh! they blasted away.

A long time later, near an island, the leatherback slowed down. "This is where I'm going," she said. "I'm going to lay my eggs here. This is where I was born, a long time ago!"

"Cool!" said Pecko, jumping off. "Thanks for the lift, you've saved me lots of time!"

The leatherback waved goodbye and swam away.

Pecko was about to carry on when he heard a voice again.

"Stuck! I'm stuck! Anyone got a rope? A pulley? Digging equipment? Hello?"

Pecko swam around the island and saw...

"An orca? What are you doing there?"

"Well," said the orca chattily, "There were no fish where I normally live, so I travelled this way. Finally, I spotted some fish and as I was chasing a shoal of them, they suddenly changed direction, so I did too, and – ta-da! – here I am."

The orca sounded quite cheerful, but Pecko could see that he had been crying.

"Don't worry," said Pecko, thinking furiously. "I'm too small to move you myself, but I'll get help!"

Pecko swam around the small island, his eyes looking here and there. At first, he couldn't see anyone who could help, but then at last, he saw a shimmer of movement in the depths of the water.

"Hello!" he shouted.

The shapes wavered and turned towards him, and Pecko saw that it was five strong dolphins.

"You called?" said one of the dolphins.

"Yes, sorry…" said Pecko nervously. "There's an orca stuck on the beach. I can't move him by myself! Please come and help?"

The dolphins quickly formed up and followed Pecko. Once they got to the orca, the dolphins tried to move him, but they couldn't.

The oldest and strongest dolphin said, "It's no good."

Pecko said, "Surely there's something we can do?

"We need a rope to make a harness that we can all pull on. As it is, we're all trying one at a time, which doesn't work!"

"I can help!" Pecko quickly took the rope out of his pack and, following the dolphin's instructions, made it into a harness that all five dolphins could fit into.

Then Pecko tied the other end of the harness to the orca's tail.

"One!" said the oldest dolphin. "Two, three, pull!"

All the dolphins pulled hard, and the orca slid down into the water, blowing a spout of water joyfully through her blowhole.

"Yay!" cheered Pecko.

"Yay!" cheered the orca.

"Yay!" cheered the smallest and youngest of the dolphins. The other dolphins just smiled and nodded as the orca said thank you to them. Then they swam off, back in their formation, soon getting lost in the depths of the water again.

The orca turned to Pecko. "They might have pulled me back into the water, but you saved me! If you hadn't heard me…" he shuddered. "Anyway! My name's Odin."

"I'm Pecko! I'm on a journey to the Arctic to see my friend, Polo."

"That's quite a long way for such a small penguin. Would you like a lift?"

"Yes, please!" said Pecko. "If it's convenient, of course!"

In no time at all, Odin was dropping Pecko off near the ice shelf of the Arctic. "Mmm, bracing!" said Odin. Orcas don't mind the cold at all.

"Brrr!" said Pecko.

He set off over the pack ice, looking for Polo. He couldn't see anyone at all!

"Hello!" A patch of snow seemed to be talking to him!

"Hello?" said Pecko.

The patch of snow moved, and Pecko saw that it was an Arctic fox, all fluffy and white as snow.

"Can I help you?" she asked politely.

"I'm looking for my friend, Polo. He's a polar bear!"

"I can help! I know Polo!" smiled the fox. "This way."

The fox led the way around a glacier and there was a neat house with a line of fish on the roof, drying out ready for the winter.

"Thank you!" said Pecko to the fox.

"My pleasure!" said the fox.

Pecko knocked on the door and listened to the footsteps that approached. It swung open and there was Polo! Polo swept Pecko into a giant polar bear hug, and Pecko felt as happy as can be.

Polo made some hot chocolate for them both and as they sipped it, Pecko told Polo about all the adventures he'd had along the way.

Polo listened thoughtfully. "I've been worrying," he said at last. "The ice is melting and there are less fish. But you have shown me that there is always something to be done. You've taught me that I need to say, 'I can help!' instead of 'somebody help me!' Thank you, Pecko!"

Pecko smiled. "We can all help!" he said. "And that will make everything better!"

Lightning Source UK Ltd.
Milton Keynes UK
UKHW050747090321
379876UK00021B/97